THE INTERMEDIATE TECHNOLOGY TRANSFER UNIT

The Intermediate Technology Transfer Unit

A Handbook on Operations

JOHN POWELL

Practical Action Publishing Ltd
25 Albert Street, Rugby,
Warwickshire, CV21 2SD, UK
www.practicalactionpublishing.com

First published in 1995
Transferred to digital printing in 2008

A catalogue record for this book is available from the British Library & Library of Congress

ISBN 978-1-85339-314-3 Paperback
ISBN 978-1-78044-438-3 Digital book

Citation: Powell, J. (1995) *Intermediate Technology Transfer Unit: A handbook on operations*, Rugby, UK: Practical Action Publishing https://doi.org/10.3362/9781780444383

Since 1974, Practical Action Publishing has published and disseminated books and information in support of international development work throughout the world. All print editions are produced and distributed via ethical and sustainable print on demand global facilities.

Practical Action Publishing is a trading name of Practical Action Publishing Ltd (Company Reg. No. 01159018 | VAT 880 9924 76). All profits are covenanted back to its parent group, Practical Action (Charity Reg. No. 247257).

The views and opinions in this publication are those of the author and do not represent those of Practical Action Publishing Ltd or its parent charity Practical Action. Reasonable efforts have been made to publish reliable data and information, but the author and publisher cannot assume responsibility for the validity of all materials or for the consequences of their use.

Typeset by Dorwyn Ltd, Rowlands Castle, Hants, UK

The manufacturer's authorised representative in the EU for product safety is Lightning Source France, 1 Av. Johannes Gutenberg, 78310 Maurepas, France. compliance@lightningsource.fr

Contents

Preface to First Edition

Many men have contributed to the evolution of the Intermediate Technology Transfer Unit from an idea in 1975 to an operating reality a decade later. The then Vice-Chancellor of the University of Science and Technology, Professor E. Bamfo Kwakye, and the Asantehene Otumfuo Nana Opuko-Ware II were instrumental in bringing the project to the attention of the Government of Ghana. Then it was Mr R. W. Kwami and Dr E. Taylor of the Ministry of Finance and Economic Planning who saw the promise of the concept and recommended the release of funds to purchase the building to house the first ITTU at Suame Magazine, Kumasi.

However the greatest contributions have been made by the two men who have served as manager of the Suame ITTU. Mr Sosthenes Buatsi was the pioneer in this role from August 1980 until October 1984. It was he who discovered what was possible and what was not possible and in doing so brought a healthy pragmatism to ITTU operations. Sosthenes Buatsi was followed by Ralph Moshage who is remembered for having added the iron foundry to the Suame ITTU and for having defined the need and created the role of the training officer.

There are many others who by their efforts have helped to promote the concept of the ITTU and to make it work. A few who must be mentioned include Dr Ben Ntim, Kevin Davis, Gordon Walker, Marcel Brosseau, Dr Francis Acquah, Edward Opare, Samuel Arthur, Frank Awuah, Solomon Adjorlolo, Gilbert Workey, S. O. Larbi, S. A. Okunor, Frank Robertson, David Wright and Bob Spencer. Mention must also be made of Ian Smilie who through the publication of his book *No Condition Permanent* brought the ITTU to the attention of a much wider public.

Finally thanks must be given to Professor Arthur Francis who by inviting the author to the University of Nairobi in 1974 sparked the original idea that grew into the ITTU. Those many friends who helped but have not been mentioned will no doubt find it in their hearts to forgive the omission.

JOHN POWELL

Kumasi
July 1986

Preface to Second Edition

When the first edition of this handbook appeared in 1986 there was only one Intermediate Technology Transfer Unit (ITTU) in full operation. By 1992, in addition to the original ITTU established by the Technology Consultancy Centre (TCC) at Suame Magazine in Kumasi, there were ITTUs in operation at Tamale, Tema, Cape Coast, Ho, Sunyani, Wa, Koforidua and Takoradi: all set in motion by the GRATIS Project since 1988. Whereas in 1986 the TCC had acquired six years of experience in ITTU operations, by the end of 1994 the TCC and GRATIS had accumulated a total of 44 ITTU-years spread over nine of the ten regions of Ghana.

This rapid expansion of the application of the ITTU concept to new geographical situations prompted a review and update of this handbook, not so much to adapt it to new circumstances as to define in more detail the function of its various grades of staff and the nature of services provided for clients of various categories. In its basic form, the ITTU has been shown to possess enough inherent flexibility to adapt to the needs of the nine regions in which it is so far established and no fundamental changes are proposed. At the same time, it has been realized that because of its unique characteristics, the functions of the ITTU and its officers, technicians and apprentices need to be spelled out in some detail if newcomers are to come into the system and assume an effective role smoothly in its operations. Thus this revised edition has taken on more of the nature of a training manual than its predecessor, although it is still hoped that it can serve a general interest need in introducing the ITTU to a wider audience.

JOHN POWELL
Nottingham
July 1995

Historical background

The concept of the ITTU evolved in the early 1970s from the interaction of the Technology Consultancy Centre (TCC) of the University of Science and Technology, Kumasi, with the artisans of Ghana's largest informal industrial area, Suame Magazine, situated in the north-western suburbs of the city. The plan for the ITTU was ready by March 1975 but more than five years of fundraising and preparations preceded the start of operations in August 1980. By that time the population of Suame Magazine, masters and apprentices, had grown to 27 000 and it was destined to top 40 000 by 1984.

Suame Magazine, like most informal first-generation engineering industries in Africa, was founded to minister to the needs of the motor car. A survey of 1971, when the population was a little over 5000 employed in 1000 separate enterprises, showed that more than half the artisans were engaged in some aspect of the repair of road transport vehicles. Trades represented included general auto-fitting, engine overhauling, panel beating (body straightening), vulcanizing (puncture repair), auto-electrics, battery doctoring and upholstery. Trades of a more general nature such as carpentry, blacksmithing and electric-arc welding were also found to focus on the repair of cars and trucks, often adapting their bodies to serve a new role as tro-tros, mammie wagons or cocoa trucks.

Although in 1971 the predominant role of the Suame artisans was in the repair of imported machinery, the beginnings of some manufacturing activity was already discernible. The repair and adaptation of truck bodies had advanced to the point where whole new wood and steel bodies were built upon old chassis and in this evolution it was difficult to draw the line where repair ended and manufacturing began. New products began to appear that made use of old vehicle wheels and axles. These included trailers for agricultural tractors and pull-along trolleys with four car wheels which gained universal popularity in the markets of Ghana for the short-range transport of goods. Electric-arc welders began to produce charcoal stoves from the sheet steel of old car bodies and extended their range of products into steel burglar-proof screens and ornamental ironwork.

1

The stirrings of manufacturing activity at Suame Magazine attracted the attention of a group of lecturers of the Faculty of Engineering of the University who decided to do what they could to encourage the trend by forming the Suame Product Development Group. Although in less than one year the activities of the Group began to be absorbed into the new and official Technology Consultancy Centre, its efforts eventually resulted in the widespread adoption of semi-automatic capstan lathes to supply the needs of the Magazine for thousands of steel bolts and nuts of many different types.

In the early 1970s, the TCC established production and training units on the university campus in an attempt to attract the interest of the artisans in new production techniques and the manufacture of new products. However, by 1975, it had been realized that the proper location for such an effort was in the heart of the informal industrial area and the basic concept of the Intermediate Technology Transfer Unit came into focus. The Suame ITTU became a reality when the Ministry of Finance and Economic Planning provided a grant of ¢500 000 to purchase a suitable building and the Canadian International Development Agency (CIDA) provided support for machines, vehicles and technical assistance amounting to Cdn $250 000. On 23 February 1981 the Minister for Industries, Science and Technology, Mr M. P. Ansah, formally commissioned the Suame ITTU. He was accompanied by his two deputy ministers, one of whom, Dr Francis Acquah, was later to play a leading role in further promoting the role of the ITTU on a national scale.

In 1982, the artisans who had been assisted by the Suame ITTU decided to form the TCC Clients Association and a year later they called upon the TCC to join them in mounting an exhibition of their products in the national capital, Accra. The Ghana Can Make It Exhibition, held at the British Council Hall in October 1983, led to a call for ITTUs to be established not only in Kumasi but in all the ten regional capitals of the country. When Dr Francis Acquah returned to the Ministry of Industries, Science and Technology as Secretary of State late in 1984 he called on the TCC to draw up a plan to achieve this objective.

The plan that the TCC presented to Dr Acquah called for the establishment of the Ghana Regional Appropriate Technology Industrial Service (GRATIS) which came into being by order of the Government of Ghana on 3 February 1987. The new project was able to make a flying start because of groundwork already done by the TCC and the ministry at Tamale in the Northern Region and at the port city of Tema in Greater Accra Region.

At Tamale, since 1982 the TCC had been working on the establishment of a second ITTU under the DAPTT Project, sponsored by the United States Agency for International Development (USAID). The Tamale ITTU was transferred from the ministry to GRATIS in August 1987 and was ready for formal commissioning by Dr Francis Acquah on 11 April 1988.

At Tema, the ministry had for some years operated a small-scale industry development and training centre that was largely inactive. The ministry had on two occasions asked the TCC to operate the facility as an ITTU and some staff training had been provided at Suame. With the advent of GRATIS, the task was accomplished and Dr Acquah commissioned the third ITTU on 8 June 1988.

This was all that could be accomplished with existing facilities and any further ITTUs needed to be started from scratch. This was expected to take some time. The project was helped forward, however, by a windfall of machine tools transferred to GRATIS from the Ministry of Local Government which had imported them from India for a project in Cape Coast. With a grant of ¢18 million from the EC Counterpart Fund, a suitable building was acquired and on 2 December 1988, Dr Francis Acquah was able to commission Ghana's fourth ITTU at Cape Coast, his home town.

By the time that Dr Acquah left the Ministry of Industries, Science and Technology late in 1989 he had seen the number of ITTUs grow from one to four and its services extended to the north and south of the country. A strong skeleton had been constructed on which to build the rest of the ITTU network. Moreover, he had put in place the GRATIS Project with the necessary resources to ensure that the task would be accomplished.

In addition to the vision, enterprise and drive of Dr Acquah, the GRATIS Project drew its strength from two powerful supporters: the EC and CIDA. The European Development Fund provided in the first phase of its support an amount of ECU 1.2 million to provide for the GRATIS headquarters organization in the form of technical assistance, vehicles, office equipment, training and a small projects fund. CIDA opted for direct support to the ITTUs and in its first phase selected Tamale, Tema, Ho and Sunyani for a five-year project amounting to Cdn $3.5 million providing technical assistance, machine tools and equipment for ITTUs and client industries, vehicles and training.

The first phase of the GRATIS Project was completed on schedule with the commissioning of the Ho ITTU on 15 August 1990 and the Sunyani ITTU on 18 December 1990. The second phase was anticipated by the opening of a pilot workshop in collaboration with the British NGO,

TRAX, to prepare the way for the ITTU in Bolgatanga. This pilot project was later relocated at Wa in the Upper West Region. During the second phase of the GRATIS Project the EU (EC) provided funding for the ITTUs in the south of Ghana, at Koforidua and Takoradi, and CIDA agreed to support the ITTUs in the north, at Wa and Bolgatanga. At the time of writing, nine of Ghana's ten regions were being served by an ITTU and the commissionings have been as follows:

1 Suame ITTU, Kumasi, Ashanti Region	*February 1981*
2 Tamale ITTU, Northern Region	*April 1988*
3 Tema ITTU, Greater Accra Region	*June 1988*
4 Cape Coast ITTU, Central Region	*December 1988*
5 Ho ITTU, Volta Region	*August 1990*
6 Sunyani ITTU, Brong-Ahafo Region	*December 1990*
7 Wa ITTU, Upper East Region	*February 1993*
8 Koforidua ITTU, Eastern Region	*November 1994*
9 Takoradi ITTU, Western Region	*January 1995*

Only one ITTU remained to be commissioned:

10 Bolgatanga ITTU, Upper East Region	*scheduled 1996*

Intermediate Technology Transfer Unit

In physical form, the ITTU consists of a group of basic engineering workshops located in the centre of an informal industrial area. The activities are as follows:

- Metal machining
- Welding, steel fabrication and sheet metalworking
- Blacksmithing
- Ferrous and non-ferrous metal casting
- Woodworking/patternmaking

Each section of the ITTU is designed to mirror an informal sector workshop in which a master is assisted usually by between four and six apprentices. Thus each section of the ITTU is headed by a skilled craftsman who works with a group of five or six apprentices. It is rare in the informal sector for a master artisan to employ another master and it is also rare for the ITTU to employ more than one technician in a section but it happens when a section is a particularly busy one.

Each section of the ITTU maintains continual manufacturing activity and sells the products it produces. While the emphasis is on promoting new manufacturing activity, the ITTU may also offer repair services where these introduce a new technology or support the work of clients who yet lack skills or facilities to accomplish the full task in their own workshops. The ITTU seeks out opportunities for the supply of new products or services and pioneers them until clients gain the skills and resources to take up the new activity. The process is a dynamic one in which the ITTU seeks to hand over activities to its clients as they become competent and continues to probe ahead to identify new products and manufacturing technologies to meet future needs and market opportunities.

By operating commercially and living to the greatest possible extent off its earned income, the ITTU tests in a meaningful way the viability of its various activities. It is therefore able to supply its clients with reliable economic data and to demonstrate in an easily recognized form where profits are to be made. The slogan, "It is profit that transfers technology!" has been adopted to stress this point. No technology that is not viable can be expected to transfer to a sector that knows the hard economic facts of life.

Many of the products introduced by the ITTU are tools and machines needed by small-scale industries located in the towns and villages of the region it serves. Some of these products may be for use in engineering industries and these include items such as metalworking tools, furnaces and power hammers but most will be for non-engineering industries such as agriculture, food processing, building construction and craft industries. The non-engineering or secondary industries provide many opportunities for the ITTU to introduce new equipment and this presents great potential for exploitation by the light engineering sector.

To develop fully the potential of the non-engineering industries of its region, the ITTU manager is assisted by a rural and women's industries extension officer (RAWIEO). This officer seeks out new opportunities for the local manufacture of tools and equipment and where they are introduced he/she sets up appropriate training programmes to ensure their effective use. In areas where long-term training is required this has led to some ITTUs acquiring non-engineering production and training sections. An example of this is the broadloom weaving section of the Tamale ITTU. In other cases, where only a few days training is required, the RAWIEO arranges short courses or workshops. It is by this means that the ITTUs have promoted beekeeping as one of their most extensive and successful projects, reaching thousands of people in all parts of Ghana.

As the title implies the RAWIEO puts much effort into promoting employment opportunities for women. This effort extends not only to traditional women's industries such as food processing, pottery and charcoal production but also to traditional mens' industries such as cloth weaving and engineering. In the Ashanti and Northern regions, the taboo excluding women from weaving has been broken by the introduction of the European broadloom and in Kumasi and Tamale more women than men present themselves for training. In the engineering field, progress has been slower but all ITTUs seek to enrol female apprentices and Tema and Tamale have succeeded in discovering some exceptionally talented female technicians who more than hold their own with their male counterparts. It is planned that the female engineers will assist the introduction of mechanical technologies into traditional womens' industries and prevent the gender change which has often accompanied attempts at mechanization in the past.

It is through its programme of rural and womens' industries that the ITTU reflects the characteristics of its region. The necessary basic engineering technologies are the same everywhere but the secondary industries of a region are derived from its natural resources and traditional skills. Thus rice, cotton and bullock farming have strongly influenced the work programme of the Tamale ITTU in contrast to the palm oil, maize and woodworking regime which largely shaped secondary industry activity in Kumasi. At Tema and Cape Coast the demands of the fishing industry have been felt while in all the southern regions there is much interest in the mechanization of the processing of cassava into its preserved forms of gari and kokonte. The Ashanti and Volta regions have strong traditions of craftsmanship in a wide range of long-established rural industries but the challenge for the Ho ITTU is to attract the talents of its native people back to the Volta Region where the fortunes of some industries have been adversely affected by emigration to other regions.

The diversity of secondary industries has meant that while the ITTUs all possess essentially the same range of engineering capabilities, their manufacturing programmes reflect a regional bias. In Tamale much effort has been devoted to the production of cotton spinning wheels and weaving looms, Cape Coast has specialized in the manufacture of winches for well sinking while the Tema ITTU foundry has served the needs of the port city's large industries for spare parts for their imported machinery. By these examples it is seen how the ITTU identifies new opportunities for local manufacturing and alerts the small-scale and informal sector to new areas of endeavour. It promotes linkages between the urban sector and the rural sector, the formal and the informal, the large-scale and the small-scale, the traditional and the modern. The ITTU works amongst the people at the grassroots, sharing the same economic and social environment, seeking with them to engineer a way forward that is self-reliant and self-sustaining. It promotes an integrated approach that seeks to encompass all those characteristics which have come to be identified with Intermediate or Appropriate Technology in the tradition of Dr E. F. Schumacher and the Intermediate Technology Development Group.

In summary it may be useful to list those features which individually and in combination make the ITTU a unique institution.

Informal sector focus

The ITTU is designed to promote grassroots industrial development by means of technology transfer to the small-scale and informal sector. It aims to extend to this disadvantaged sector some of the benefits enjoyed by the large-scale formal sector industries such as consultancy services, training and access to imported plant and machinery.

Free-of-charge services

The ITTU provides three basic services free of charge: information, advice and training. The informal sector cannot afford to pay for consultancy services and even if it could, it probably would not understand the concept or appreciate its benefits. For this reason the ITTU does not charge for information and advice given to clients in both technical and commercial areas. No charge is made for training because to a large extent the trainee repays the ITTU by his labour and this enables the ITTU to remove any barrier that a charge might impose.

Self-financing

The ITTU is designed to be self-financing in terms of recurrent expenditure. To date all operational ITTUs have been able to generate sufficient income to cover all expenses except for the salaries of 10–12 core staff on government subvention. In practice this has meant that they are self-financing to an extent of 75–90 per cent. Thus the earned income multiplies the government input from 4 to 10 times and by this same ratio amplifies the impact of the government's investment. There is no discernible reason why an ITTU should not be 100 per cent self-financing in respect of recurrent expenditure and several ITTUs have achieved this over limited periods of time. However it is unlikely that an ITTU would ever be able to recover the capital costs incurred in its establishment.

No ab-initio training

The ITTU is not a technical training institute in any formal sense. It does not train beginners. It rather provides an in-service training programme for the informal sector. All the trainees in its engineering sections come to the ITTU either with practical skills acquired from a master artisan or some theoretical knowledge acquired at a technical school or polytechnic.

Training for self-employment

The main objective of the ITTU's training programme is to establish the trainee in self-employment. Training is supplemented by a range of services to help establish the new enterprise and sustain it for several years.

Material assistance

The ITTU assists its trainees and clients to acquire the machine tools and manufacturing equipment that they need to establish their own workshops by importing and selling on cash and credit terms. The cost of this service is often minimized by the importance of good used machine tools rather than new machines that often cost more than the clients could afford.

Services of the ITTU

Purpose of the ITTU

The purpose of the ITTU, as its name implies, is technology transfer. The focus of its activities is the small-scale and informal industrial sector. Thus the ITTU attempts to transfer technology to small-scale industrialists and entrepreneurs who seek to introduce new products or to employ a new manufacturing technique. The technology is new in the local sense that it is not widely known and used by the small-scale and informal industries in the immediate locality. The ITTU selects, adapts or develops an appropriate technology that is more advanced than the technology generally employed but not so advanced that it is beyond the skill or resources of the ITTU's clients. In other words, the ITTU attempts to provide the client with a technology for which he is ready or can be made ready. An attempt is made to define the next logical step on the technology ladder and to help the client to ascend it.

The ITTU is still a relatively new concept. Thus its mode of operation and range of services are still evolving. The ITTU is essentially a flexible organization that should be able to adapt its services to meet the varied needs of its clients. An innovative and imaginative approach to the task will inevitably generate new services as time goes on. Thus no list of services could be complete and final. However, the basic services are now well defined and a listing can be useful to those new to ITTU operations. It should by no means preclude the introduction of new services but should rather provide a seed bed from which other ideas for services can grow.

The ITTU does not seek to dictate to its clients. It attempts to present itself to them as a source of help in the development of their business. It listens to the client's problems and tries to devise a strategy to assist him in realizing his own chosen ambitions. Some clients already know what they want to achieve. Many others become inspired by seeing what others are already doing. To attract these, the ITTU begins by demonstrating some new ideas for productive ventures.

Demonstration

The workshops of the ITTU are in constant production and either employ new manufacturing techniques or make a product that is new to local industry. Thus they demonstrate to potential clients new industrial activities which they might take up in their own workshops.

It is not only technology that is demonstrated. The whole idea of continual, methodical and disciplined manufacturing may be new to the informal industrialist. He may be ready to learn much from the management procedures used at the ITTU such as timekeeping and attendance records, safety precautions, stock control, production planning and costing. However these procedures should also be simple and understandable to the client or their significance will be lost. They should not be stressed or perhaps not even mentioned in the early stages, but their demonstration is important. The aim is to inspire the client with the idea that here is a modern industry such as he himself might own and manage with a little effort and a little luck.

The client will be drawn to the hardware aspect of the technology and the economic benefits that it can confer. The software aspects, the management methods and organization, will enter his field of vision as he becomes more involved in a project. The ITTU must, however, offer a package of both hardware and software and the first impact is made by providing a demonstration. This effect is enhanced by publicity and by making the ITTU workshops accessible, if not to all casual observers, at least to all serious enquirers.

The informal industrialist may feel that he is locked into his environment. By showing him something better in the midst of squalor and disorder the ITTU offers hope of advancement. The ITTU must present a dream but a dream that can come true.

To achieve the best possible demonstration effect the ITTU workshops must always present a neat, tidy, clean and efficient appearance. Safety procedures must be carefully observed. Machines must be regularly serviced and repaired promptly if they break down. All machines should be operating and all workers should be busy. Nothing destroys the desired effect as much as idle hands and the sight of machines that have obviously been broken-down for a long time. Much can be gained from paint applied regularly to both machines and buildings. This is an aspect which is often totally neglected in informal sector workshops and will therefore create all the more impact. Also, the surroundings of the buildings should be kept clear of vehicle carcases and piles of scrap steel and the ground should be

11

regularly weeded and levelled. Although the ITTU is situated in the midst of an informal industrial area, the immediate impression should be created that here is something that is different and better. The ITTU should excite curiosity and attract attention. This is its demonstration effect.

Retail store

One important means of attracting clients to the ITTU is by the operation of a retail store. The retail store sells products manufactured in the ITTU and client workshops. For example at the Suame ITTU the following products have been sold:

- Steel bolts and nuts of many different types
- Hoes, cutlasses and other farming implements
- Bars of metal recast from scrap including brass and bronze, aluminium and cast iron
- Corn mill and pepper mill grinding plates
- Charcoal and sawdust burning stoves
- Soap

The retail store can also extend its range of goods by stocking the products of clients. At times when soap was scarce the Suame ITTU attracted much custom by selling laundry soap produced by Kwamotech Industries Limited, a client of the TCC. At other ITTUs the sale of honey has served the same purpose.

Informal industrialists calling at the retail store may see products that they would like to produce in their own workshops. This is the primary purpose of the retail store. Its operation can also increase the income of the ITTU and that of clients whose products are sold there. But even if the additional income generated does not cover the cost of the retail store it is still worth retaining for its demonstration effect.

Information and advice

The supply of information and advice is a service that the ITTU can provide to a client from the first visit and at all times thereafter. ITTU staff should provide clients with all relevant information on technical and other related aspects of the work. Advice and information should be freely given in every sense. In Ghana, where information is often withheld and denied to the enquirer, the free supply of information can do much to encourage the client and win his confidence.

ITTU staff must keep themselves well informed on both practical and theoretical aspects of the technologies they use. Efforts must be made to ensure that all information given out is accurate and reliable. This process can be supported by a small reference library of technical books to which reference should always be made in case of doubt. Clients may be invited to study the books themselves but it is not advisable to allow the books to be removed from the library.

Where technical information is of common interest to a number of clients, it is useful to produce printed data sheets. For example the TCC prepared data sheets on screw threads of all standard types (BSW, BSF, UNC, UNF, metric), on standard hexagons for bolts and nuts and on capstan lathe collets. GRATIS has extended this theme to include the production of small handbooks on topics such as the operation of winches, beekeeping and batik cloth printing.

Advice given to clients should always be soundly based on past experience. It is better to give no advice at all than ill-considered advice. If the advice given is later found to be disadvantageous the trust of the client will be lost. Nevertheless the range of operations of the ITTU and the knowledge of the problems of past clients constitutes a considerable store of experience on which much advice can be soundly based. It should be remembered that such knowledge and experience is stored in many heads and other staff and old clients may often be consulted to advantage.

Training

When a client wants to take up a technology demonstrated at the ITTU he can be offered a period of practical on-the-job training for either himself or one or more of his employees or apprentices. Those undergoing on-the-job training at an ITTU workshop fall into two categories: visiting apprentices and ITTU apprentices.

These categories are discussed fully in the section on staffing of the ITTU. Essentially visiting apprentices are short-term and unpaid and ITTU apprentices are longer term and paid. Most clients, who are already masters with their own workshops, will be visiting apprentices whereas most younger men with ambitions to own their own workshops in the future will attend the ITTU for a few years as ITTU apprentices.

Training at an ITTU is always free in the sense that no training fee or apprenticeship fee is charged. The visiting apprentice contributes some of his labour to the ITTU and the ITTU apprentice is paid for his labour. In general the work output covers the cost of training and the sale of the products of the ITTU workshops renders the ITTU either wholly or largely self-financing.

The ITTU may collaborate with universities, polytechnics and technical institutes to provide on-the-job training during vacation periods. These attachments are normally as visiting apprentices. Such occasions should be regarded as opportunities to interest young people with formal education in the benefits of self-employment. The focus of the training programme must remain the transfer of technology and the generation of new private enterprises.

Hire of manufacturing facilities

When a client or his employees have been trained in a new technique he will seek to establish the new activity at his own workshop. Often, however, he will lack one or more facility in the form of a machine tool or process plant. In such cases he may be allowed to hire time on a suitable facility at the ITTU workshop. For this service he will pay an appropriate fee determined on an hourly or daily basis.

Clients may hire time on ITTU facilities provided that the ITTU's own manufacturing activities are not unduly adversely affected. The needs of other clients wishing to use the facility must also be considered. It is however useful to provide the ITTU with a degree of excess capacity as this service is a most valuable one for many clients and a means of boosting their incomes to generate capital for the purchase of their own facilities.

Where a machine can be spared and where a client can make full-time use of the machine, it may sometimes be removed to the client's workshop. However, this is really only desirable where the client has expressed the intention of eventually purchasing the machine. If the machine is not sold its recovery may prove to be difficult. The long-term hiring of machinery should be avoided. The hiring of a machine should be regarded as a step towards the client acquiring his own machine.

Sale of manufacturing facilities

The ITTU should make every effort to ensure that all serious clients who have followed the advice given and have prepared themselves well to use a new technology should be supplied with the necessary manufacturing facilities in their own workshops and in their own possession. To do this the ITTU must be in a position to sell to the client the appropriate machine tool or process plant.

Informal industries in Ghana are in dire need of such basic facilities as centre lathes and electric welding plants and numerous other items that are commonplace in workshops in industrialized countries. Such facilities are in great demand. The ITTU will never possess the resources to meet the demand as fully as a trading organization might which had no other concern but to sell plant and equipment. Thus the ITTU must adopt a soft approach to this aspect of its work and supply machines and equipment only to selected clients who have proved their interest in adopting new technology. This implies that the ITTU will not offer machines for general sale to people who will operate them in the old ways of informal industries. Client selection is a topic dealt with at length in another section which should be studied in association with the present section.

The ITTU acquires machine tools and equipment for resale through a programme which is financed by foreign aid donors. Much effort must be expended on attracting the interest of donors and keeping them informed of progress and developments.

Most of the machine tools imported by the TCC between 1973 and 1986 were used machines. The foreign donor agencies and especially the Government of Ghana are often prejudiced against the importation of used machines. Attempts should be continued to educate them on the rationale behind the policy. The matter is explained at length in another section.

The selling prices of machine tools and equipment are usually determined by the donor agency with the approval of the Government of Ghana. The prices are usually based on the CIF Tema prices in foreign exchange converted to cedis at the prevailing rate of exchange. In the late 1970s and early 1980s this procedure led to very low prices which implied a high degree of subsidy on the part of the donor. Now the pendulum has swung the other way and prices seem to be very high and beyond the reach of many small industrialists. The ITTU must strive to achieve a balance and use its influence to moderate prices as much as possible.

Machine tool installation and repair

The ITTU can use its senior technical staff to assist clients with the installation and repair of machine tools and other manufacturing facilities. This service is available both for machines sold by the ITTU and for machines acquired by the client independently. The aim should be to encourage the small-scale industrialists to get the best out of their machines by means of correct installation, routine maintenance and repair. To this end the opportunities to provide these services should also be taken as opportunities to educate the client in these matters. These services must be charged at the appropriate rates for the labour and materials employed.

Subcontracting production orders

The workshops of the ITTU produce and sell and become known in the local markets as suppliers of certain commodities. Hence orders are placed with the ITTU by users and retailers of its products. As clients prepare themselves to take over the production of an ITTU product they acquire the capability to make part or the whole of that product. Thus it becomes possible to share the work with the client. This is done by subcontracting the work to the client.

Subcontracting may mean that part of the manufacturing activity is undertaken by the client and the product is finished by the ITTU or it may mean that the client does all the work and supplies the ITTU with finished products. Eventually, when the client and his staff have been trained and the necessary manufacturing facilities have been acquired, the client will be able to take over the production and marketing of the product. Until that point is reached he can be fed with work during the preparatory period by orders subcontracted from the ITTU.

Subcontracting can also work the other way round: from the client to the ITTU. Sometimes a client may receive an order for a product which he cannot completely manufacture in his own workshop. He may then turn to the ITTU to carry out one or two manufacturing operations and supply him with components or part-finished products. This fruitful collaboration supplies the client's need and provides income for the ITTU.

Clients should also be encouraged to subcontract work to one another. Although they are often reluctant to share the work in this way it is a means of extending the range and volume of work done to everyone's benefit.

Sale of small tools

Most new technologies transferred by the ITTU need to be sustained for some time by the supply of small tools and machine accessories. Such tools include, for example, taps and dies, milling cutters, capstan lathe collets and other accessories, etc. Other more commonly used small tools such as lathe tool bits, drill bits and grinding wheels are already obtainable from traders who bring them from neighbouring African states and from Europe. In the course of time, as the demand increases and becomes perceived, the other small tools will also be traded. However the ITTU moves ahead of the field and must endeavour to maintain the supply until it is supplied from other sources.

The funding for the small tools stock, like that for machine tools, comes from foreign donor agencies. As all the small tooling is new there is in this case no prejudice against the importation of used goods to be overcome.

If stocks are sufficient, small tools can be sold to all comers from the ITTU retail store. However at most times it is likely that stocks are limited and then sales must be restricted to selected clients who are using technologies and machines supplied by the ITTU. The ITTU has a primary responsibility to those clients who are pioneering new products and new manufacturing techniques. These must be sustained until they succeed and others are attracted to copy them. The copying of successful pilot projects is the means by which a wide-scale spread effect is achieved.

Other technical services

Since the ITTU possesses a wide range of machine tools and manufacturing facilities it can provide a range of technical services to support the operations of its clients. Perhaps the most widely used is a tool resharpening service. This facility is provided by a tool and cutter grinding machine which can resharpen thread-cutting dies and milling cutters as well as other cutting tools such as drill bits and thread taps.

Another very useful service for numerous clients can be provided by the ITTU foundry. Suitable scrap metal of ferrous or non-ferrous varieties can be melted down and recast into bars and tubes for clients who wish to machine them into bearing bushings, valve seatings, cylinder liners and other components for the repair of machinery or the manufacture of new products.

Yet another popular service provided by the Suame ITTU is the flame-cutting of thick steel plates. This service is helpful to clients who want to use parts cut from heavy gauge plates but lack the ability to cut them. The widely-used method for cutting steel plates in informal industries by hand-chiselling is limited to plates of about ½ inch thickness.

Experience, ingenuity and demand from clients will increase the range of technical services that the ITTU can provide. These services carry a charge which adds to the income of the ITTU. Thus both the client and the ITTU benefit, and effort should be applied in a search for other technical services which the ITTU can provide to the community at large.

Letting of workshop accommodation

The ITTU should seek to let surplus workshop space to clients who lack adequate accommodation in which to start a new technological venture. This facility has the following advantages:

- It helps the client to solve his accommodation problem
- It fills all available space at the ITTU
- It increases the range of activity at the ITTU and increases the demonstration effect
- It brings income to the ITTU

However clients should always be encouraged to seek their own permanent workshop accommodation. To this end the lease of workshop space at the ITTU should be for three years with a one-year extension and then old clients should move out to enable new clients to benefit in the same way.

ITTU managers should seek ways of raising finance to increase the amount of workshop accommodation available for letting to clients. The construction of blocks of small workshops for letting on a permanent basis should be considered. In this way the ITTU might generate its own mini industrial estate and meet a pressing need as well as set a good example in the planned development of informal industrial areas. In locations where the basic infrastructure is lacking the ITTU can provide three-phase power and water supplies to its leased workshops as an extension of its own facilities.

Supply of raw materials

Informal industries base most of their manufacturing activities on the re-use of scrap metal from abandoned vehicles and industrial machinery. The activities of the ITTU should seek ways of developing and extending this trend and the introduction of iron casting is by far the biggest step to be taken in this direction. The introduction of flame-cutting, powered hacksawing and milling and shaping machines can also make a useful contribution to the recycling of scrap steel.

However scrap metal cannot meet all the needs for raw materials in the informal sector. Some imported steel in the form of sheets, bars and tubes is essential if the range of products is to grow and new manufacturing techniques are to be fully utilized for the benefit of the local economy. The ITTU can help by seeking to import materials for its own use and for resale to clients. Another useful approach to this problem is to supply foreign-funded aid projects with finished products in return for the imported raw

18

materials needed to produce them. Labour and overhead costs are paid for by the supply of additional materials which can then be used to manufacture products to supply other local customers who do not have access to foreign exchange. Several major funding agencies have shown a willingness to operate in this way. It represents a more cost-effective use of foreign exchange than the straight importation of finished products that can instead be made in Ghana by the ITTU or its clients.

Financial aid

The ITTU's business is technology. It is certainly not a bank or a financial institution. However an integrated approach to small industry development would be incomplete without some policy and service related to meeting the needs of clients for raising funds for capital investment.

The ITTU does not possess the financial resources to provide loans or credit facilities to clients on a regular basis. Some of its services involve an element of subsidy and some machine tools may be sold to clients on hire purchase terms. This is the extent of its direct financial involvement. However the ITTU can help clients to obtain financial support by other means.

When a client wishes to take up a productive activity which is already established at the ITTU, it is possible to draw up a feasibility study and a cash flow projection based on actual operating experience. These documents, backed by the recommendation of the ITTU, can be submitted by a client to his bankers in order to obtain a loan to finance the new business or the new extension to an existing business. This support for clients has often been effective in the past and should continue to be effective in the future provided that care is taken to ensure that feasibility studies are undertaken with care and that the client to be helped has proved himself to be a reliable and trustworthy person.

The ITTU must take pains to establish and maintain its reputation with banks and governmental financial agencies as a reliable and objective adviser in these matters. Only one or two serious defaulters could destroy the ITTU's credibility. Thus it must be realized that not all clients are worthy of recommendation and not all their projects are economically viable. As in all other aspects of ITTU work the temptation to help all clients equally must be resisted and judgment must be used in recommending clients to third parties. That being the case the support of the ITTU can be a powerful factor in winning financial backing for clients' projects.

Banks and governmental credit schemes are not the only sources of finance for industrial development. Indeed in the informal sector they do not feature prominently in the normal course of events. Most informal industrialists turn first to a wealthy relative or a friend working overseas who can perhaps send him a machine tool as a loan to be repaid when the friend returns to Ghana. As a last resort there are moneylenders but most clients would be well advised to avoid these as their interest rates are too high for the financing of industrial ventures. The client should be encouraged to seek family financing of his new venture. Most small businesses are family concerns and represent the normal way of life in local society. Thus the rich parent or aunt or uncle should be approached before the bank manager and here again the recommendation of the ITTU can be effective support for the client in gaining the loan he needs.

If the client seeks advice on his financial affairs he should be counselled to observe discipline in all aspects of his business. He should be encouraged to keep accounts of all income and expenditure and to cost his products and services systematically. Above all he should be advised to attempt to accumulate profits within the business in order to finance future developments. Premature profit-taking is the cause of the collapse of many businesses. New clients should be advised to reinvest all profits for an initial period of perhaps five years. Clients should be advised to operate a bank account and to pass all income through the account to enable the bank to know the turnover of the business. This is of great help in raising a loan.

Summary of services of the ITTU

The following services have been discussed in this section.

1. Demonstration
2. Retail store
3. Information and advice
4. Training
5. Hire of manufacturing facilities
6. Sale of manufacturing facilities
7. Machine tool installation and repair
8. Subcontracting production orders
9. Sale of small tools
10. Other technical services
11. Letting of workshop accommodation
12. Supply of raw materials
13. Financial aid

The list is quite extensive and represents an integrated approach to small industry development. However the list is not exhaustive and, as mentioned already, ITTU managers and senior staff should be flexible in their approach and ready to grasp new opportunities to serve their clients in new, imaginative and innovative ways.

Staffing of the ITTU

Purpose of the ITTU

The purpose of the ITTU is technology transfer. This it does by several means which are detailed elsewhere. However one of the most important means is by providing on-the-job practical training in manufacturing processes.

The various workshops of the ITTU carry on continual production activities to demonstrate new manufacturing methods or the manufacture of new products. The products and processes are new in the sense that they are not widely found in the local small-scale industries such as inhabit the informal industrial areas surrounding the ITTU. Artisans taking an interest in adopting a new technology are invited to take advantage of the training facilities offered by the ITTU. Thus the self-employed artisan or his workers or apprentices can attend the ITTU for a period during which they undergo an on-the-job training experience that is designed to equip them to adopt the new technology in their own workshops. While undergoing training the trainees can be employed in two categories. These can be termed:

- Visiting apprentices
- ITTU apprentices.

Employment and training in these two categories will be discussed later in this section.

Staff grades in the ITTU

The ITTU is unusual in that it employs several different grades of staff and trainees. This reflects its unique function in relation to its small-scale and informal sector clientele as well as its funding support base.

The ITTU network was established by a project of the Government of Ghana; the government provides a subvention which pays the salaries and allowances of 10 or 12 core staff. The officers whose salaries are normally provided by subvention are as follows:

1 Manager
2 Assistant manager
3 Workshop supervisor
4 Training officer
5 Rural and women's industries extension officer
6 Accounting officer
7 Four or five engineering section leaders

All other permanent staff of the ITTU are paid from income earned from its operations or provided by donor agencies in support of specific projects. In general, all engineering staff are paid from earned income while staff engaged on rural and womens' industries projects are paid from earned income and sometimes by donor agency funding.

The grades of staff who are paid salaries and allowances by the ITTU from non-governmental funds include the following:

1 Leading artisans and other technical personnel employed in engineering sections
2 Typists and accounts clerks
3 Drivers, drivers' mates and labourers
4 Security officer, watchmen and security guards
5 Section leaders on non-engineering sections
6 Other rural and women's industries extension staff e.g. beekeeping instructor

ITTU technical apprentices are not paid a salary but a training allowance. This is also funded out of the earned income of the ITTU. Visiting apprentices are not paid any salary or allowance by the ITTU but they may partake of the free lunch provided for all ITTU staff and this is funded out of earned income. Trainees in non-engineering sections who may attend the ITTU for up to two years, e.g. to learn broadloom weaving, may be paid a training allowance funded from earned income or from donor support. Donor agencies sometimes support training programmes at the ITTU for rural artisans who attend for a few weeks or a few months. In such cases travelling, accommodation and boarding expenses are provided from the donor support.

Production and training

All the ITTU workshops are production workshops and sell their products or services on the local market. The income so derived pays most or all of the costs of operating the ITTU. However the ITTU is not a commercial operation in the full sense of the term. In the first place it seeks not to compete with its clients but passes on orders to clients when they have gained the skills and capacity to take over the work. In the second place it does not employ permanent shop-floor workers. It encourages its apprentices to leave when they have gained a viable level of skill. This turnover of production staff tends to lower quality and output but is a necessary side effect of technology transfer, the primary role of the ITTU. The effect is mitigated by the high level of skill of the section leaders and the effectiveness of the training programme under the guidance of the training officer.

The quantity and quality of output of the ITTU is also ensured by the apprentices who already possess basic skills when they join. The ITTU does not train beginners. As stated previously, the trainees have in the main already completed an informal apprenticeship or a course of technical training. Both masters and men come to the ITTU to learn new skills which they could not acquire elsewhere.

Visiting apprentices

Visiting apprentices are not paid any salary or allowance by the ITTU. They are trainees sent to the ITTU by their masters or by a training institute, such as the NVTI, to gain practical experience and on-the-job training in a new skill. While training they are supported by their masters, parents or guardians.

The period of training of visiting apprentices is usually short. It can be a few weeks or a few months. Included in this category may be engineering students from a university or polytechnic undergoing vacation training. Many masters also attend the ITTU as visiting apprentices. Their attendance may be part-time and they may only need to be taught to operate one or two machines or carry out one process. If they bring their own work to be done at the ITTU they pay a hire charge for the machines and tools used. This process can in some cases graduate to the extensive or even full-time hire of equipment or workshop space at the ITTU.

Visiting apprentices will usually be in a minority at the ITTU. Except at school vacation times when a larger group may attend, visiting apprentices tend to come at random times and for indefinite periods. Thus they can not be relied upon to undertake the basic production of the ITTU workshops. This is mainly the concern of the ITTU apprentices.

ITTU apprentices

There are many young men and some young women who have served an apprenticeship in an informal sector machine shop or fitting shop or have completed a course at a polytechnic but yet lack certain skills, either because the master lacked the relevant skills or because he or the polytechnic lacked the necessary equipment. These young people come to the ITTU seeking to upgrade and consolidate their skills to fit them for self-employment or to increase their value to future employers. However they can not afford to undergo formal training because they lack the means to support themselves. To meet this demand and also to provide the ITTU with its basic shop-floor labour force these young men and women are employed as ITTU apprentices.

The danger with this category of trainee/employee is that they may come to regard the ITTU as a permanent employer. If this idea becomes prevalent in the workshops the ITTU's purpose of technology transfer can be damaged or even defeated. To offset this tendency the ITTU apprentices are employed on one-year contracts. Contracts are renewed subject to:

- The training officer certifying that progress in acquiring technical skills has been satisfactory
- The workshop supervisor certifying that conduct has been satisfactory.

Technical skills are assessed by means of a practical trade test. Conduct assessment is based upon time-keeping and attendance records and the appraisal of supervisory staff.

Rates of pay for ITTU apprentices are adjusted to be above the official minimum wage. Twenty-one days paid leave is allowed within each one-year contract and a leave travelling allowance is paid to defray the cost of travelling to and from the home town.

Employment generation

Apprentices must be encouraged by all means not to regard their employment at the ITTU as permanent but as a stepping stone to something better. At most the ITTU can provide 20 or 25 work places. However, if five masters pass through the ITTU each year and, for example, 10 apprentices move on to employment and self-employment, the operations of the ITTU could be generating as many as 30 new jobs every year from its training programme alone. Every effort must be made to hold to the original aims and objectives of the ITTU. This may involve resisting pressures and temptations to provide long-term employment and security for a few at the expense of lost opportunities for the many. This is a fundamental aspect of the ITTU concept. Managers, workshop supervisors, section leaders and training officers should keep it always in mind.

Duties of principal officers

The ITTU is unlike any other institution that preceded it. It has some of the features of a commercial manufacturing enterprise and other characteristics which might be shared with a technical school or a research institute. Yet all its activities are moulded to harmonize with an informal industrial setting. It is an interface between the formal and the informal and a bridge between education and commerce, one more link between the government and the people. Such a hybrid organization makes great demands upon the skills of all its supervisory staff. They must always be searching for a middle way between conflicting interests and seeking to create harmony amongst apparently irreconcilable factors.

The interests of the clients must come first and yet the ITTU must earn enough income to survive. The ITTU stands in the front line of the international aid movement and yet its role is not to provide hand-outs but to educate and develop the potential of its informal industrialist clients. The ITTU brings tools which the people need but they are to be bought at fair prices, not stolen.

It might seem that the management of an ITTU is an almost impossible task. No doubt any one manager is liable to be biased one way or another or he may even swing between extremes especially during his early days of responsibility. There is only one sure guide and that is always to seek the path which seems to give the most direct route to the transfer of technology into private workshops. This primary role of the ITTU must guide and shape all day-to-day decisions. The ITTU is designed to overcome the isolation of the formal sector from the informal sector. The greatest danger lies in becoming yet another formal enclave in the informal environment of the Third World. It is tempting to concentrate resources on research and development activities or to run the workshops to earn a large profit or to provide secure long-term employment for a few friends or relatives. However all these temptations must be resisted.

The success of the undertaking can be gauged by the traffic across the interface. People, ideas and equipment must all be constantly coming into and going out of the ITTU. It is not only a matter of quantity but of quality too. Once again a mature judgement must be made. Overall it would seem that ITTU senior officers need a clear sense of objectivity combined with an earnest desire to help the disadvantaged not by charity but by education.

The ITTU manager

While it is the duty of the ITTU manager to set the pattern of activities for the whole organization, his own main focus must be outwards to his clientele. He must schedule his duties to provide time for meetings with clients old and new, in their private workshops and at the ITTU. He must resist the tendency to be drawn too deeply into the day-to-day administration of the workshops to the extent that he becomes preoccupied with the problems of raw materials supply and the cash flow situation. He may model himself on the captain of a cruise liner who devotes himself to entertaining the passengers while his technical officers man the navigating bridge and the engine room. The chief technician will run the workshops and deal with the problems of labour and materials. The training officer will ensure that apprentices receive a structured course of training and the accounting officer will keep track of income and expenditure. The manager will ensure that all subordinate staff are competent and then content himself with monitoring their activities.

The ITTU can provide a wide range of services and new services may become apparent from time to time. It is the manager's task to determine which services should be supplied to each client. To do this he must monitor the progress of clients and assess both their degrees of personal commitment and their needs for further assistance. In arriving at all such decisions he consults his senior assistants but the final decision to allocate ITTU resources is his.

The ITTU manager meets at intervals of two or three months with his local regional advisory board or management board who lay down policy guidelines and assist in securing additional resources for the ITTU. He meets weekly with his senior staff committee comprising the following members:

Manager	Chairman
Chief technician	Member
Training officer	Member
Section leaders	Members
RAWIE officer	Member

The manager is responsible for preparing project proposals for the extension of ITTU activities. He also produces case studies of projects when

28

suitable opportunities arise. An annual report records all major activities pursued during the year and details all services provided to major clients.

Where a regional management board has been established the ITTU manager reports to the chairman of the board. However until the establishment of management boards has been accomplished the ITTU manager reports to the director of the parent organisation, TCC or GRATIS. Ultimately it will be the management board which appoints established staff to the ITTU and the ITTU manager will be a member of the board. Casual staff may be appointed by the manager in consultation with the chief technician and the section leader concerned.

The focus of the ITTU manager is outwards towards the community at large. He represents the ITTU with local authorities and institutions, he negotiates collaborative projects with local development agencies, Ghanian and foreign, and he maintains a dialogue with the ITTU Clients Association. He also receives visitors to the ITTU, from VIPs and officers of collaborating agencies to clients and customers of the ITTU.

A professional engineer with the rank of senior technical/technical officer, the ITTU manager has overall responsibility for the engineering activities of the ITTU. He is concerned with the extension and upgrading of the manufacturing facilities of the ITTU and the enhancement of the quality of its products and services. In particular he is concerned with all aspects of technical training provided by the ITTU and especially with the ITTU technical apprentice programme.

The ITTU manager is responsible for the maintenance and extension of the Rural and Womens' Industries Programme of the ITTU. His aim is to ensure that the products of the engineering workshops of the ITTU and its clients are widely distributed to upgrade rural and womens' industries throughout the region. Following the advice of the regional board, collaborating agencies, clients and staff of the ITTU, he is responsible for drawing up project proposals and soliciting and allocating resources to promote an ever-widening programme.

The manager is responsible for the financial control of the ITTU. He prepares annual estimates of the recurrent and capital funding needs of the ITTU, ensures that proper financial records are kept, provides monthly and annual accounts of income and expenditure and co-operates in annual and ad-hoc auditing exercises.

Disciplinary control of the ITTU is also the manager's responsibility. He maintains a security force under a suitably qualified officer to maintain security of all ITTU property at the main compound and at all residential properties owned or leased by the ITTU 24 hours a day and 365 days a year. He maintains stores for tools, raw materials and finished products under the control of a suitably qualified storekeeper and ensures that all standing stores control regulations are adhered to. When a disciplinary offence is committed the ITTU manager is responsible for ensuring that a full investigation is undertaken and that standing disciplinary procedures and regulations are followed.

The manager is responsible for the safety of all personnel employed by the ITTU or under training on its premises. He maintains facilities for fire-fighting and for administering first aid and he inculcates safety consciousness in all staff and trainees.

The task of managing an ITTU is not an easy one and it calls for a special blend of skills, imagination and dedication. To a large extent the manager can mould his activities to suit his abilities and interests. Technology transfer is far from being an exact science. Much still remains to be discovered and many ideas remain to be tested in practice. All the difficulties will be surmounted by a manager who is well prepared and who is, above all, interested in people and their development. The task is both a challenge and an opportunity.

Assistant manager

The assistant manager reports to the manager and deputizes for him in his absence. He is a professional engineer and has the status of an assistant technical/technical officer. The assistant manager undertakes on behalf of the manager that part of the manager's duties which is delegated to him.

Both the manager and the assistant manager of an ITTU are professional engineers and when both officers are at post the opportunity is afforded for one or both to devote much time to engineering work. This enables attention to be paid to the design and development of new products, the introduction of new manufacturing processes and the upgrading of training programmes.

The post of assistant manager allows the incumbent to become familiar with all aspects of ITTU management and may equip the person to be a full manager in the future.

The workshop supervisor (chief technician)

The role of the workshop supervisor, though more limited in scope, is in its way as challenging as the role of the manager. He must also balance the needs to transfer technology and provide training against the needs to earn income and to introduce new technologies. He is in charge of all production workshops of the ITTU and oversees the work of all its technical staff and production operatives. He must ensure that the ITTU always presents an impression of bustling activity, that everyone is busy and employed in a disciplined setting with an obvious sense of purpose. The role has some of the attributes of a stage manager for the activities of the ITTU are an exposed and ever-present example of how informal sector workshops might be run.

When they have no orders to execute, informal industrialists sit and wait for a new order to be brought to them. They have ample opportunity to watch the activities of the ITTU and the sight should excite their curiosity and invite their attention. The task of proposing improvements is that of the workshop supervisor.

The workshop supervisor reports to the manager and is responsible to him for the operations of all the engineering workshops of the ITTU:

- Metalworking machine shop
- Welding and steel fabrication shop
- Blacksmithing workshop
- Foundry
- Woodworking and pattern-making shop

The workshop supervisor has the status of a principal technician/chief technician in the universities of Ghana and is the senior engineering technician of the ITTU. He is responsible for setting the standards of craftsmanship achieved in the engineering workshops of the ITTU. He is the officer responsible for the final inspection of all products manufactured at the ITTU and serves as the quality control officer of the ITTU.

As staff member responsible for the work programmes of the engineering workshops, the workshop supervisor determines the scheduling of jobs, the sequence of operations and the time to completion, and he advises the manager and customers of final delivery dates. He assesses the material content and labour effort needed to complete each job and supplies this information to the accounting officer to enable a costing to be made.

The workshop supervisor collaborates with the training officer to ensure that the ITTU technical apprentices and visiting apprentices are fully involved in the manufacturing activities and receive varied practical on-the-job experience consonant with their skill level and training needs. He also co-ordinates the work of the engineering workshops under their section leaders to ensure the smooth flow of work within sections and between sections. His aim is to maximize work throughput and income generation while maintaining the quality of products and training programmes.

The workshop supervisor is responsible both for the repair and maintenance of all machines and tools in the ITTU workshops — he advises the manager on the needs for new facilities and the replacement of losses and breakages — and for discipline in the workshops. He ensures that correct and safe working procedures are followed and that safety equipment is used and repaired/replaced as necessary.

It is largely the responsibility of the workshop supervisor to determine to what extent time may be hired to clients on the machine tools and manufacturing facilities of the ITTU workshops. He must rise above the feeling that this matter is a conflict between the needs of the ITTU and the needs of clients. Although it might be said that his focus is inwards, while that of the manager is outwards, he must spend some time meeting clients and discussing their needs so that he can form a balanced view of the allocation of ITTU resources. He must appreciate that the success of his stewardship of the workshops is measured largely in terms of the number of apprentices trained and the number of clients assisted.

Above all the workshop supervisor must believe that the most valuable output of the ITTU is the technology it transfers to private workshops and the ultimate objective of every production section is to assist its clients to drive it out of the market. There are a million new products to be made and a million new processes to be demonstrated. The ITTU must always be one rung ahead of its clients on the ladder of technology. Climbing steadily up that ladder is the responsibility of the workshop supervisor.

The training officer

In the early days of the Suame ITTU, it was believed that providing on-the-job training in the ITTU workshops was sufficient. The ITTU does not train beginners. All its apprentices have completed at least an informal apprenticeship and have already acquired the basic skills of their trade. The ITTU experience was intended to expose them to new applications of their skills and to add a few new skills required in the mastering of a new product or process. However experience has shown that a more structured learning experience is necessary. Firstly, there are some gaps in the skills acquired in an informal apprenticeship and these need to be identified and filled by appropriate instruction. Then there is a need to ensure constant progress on the part of apprentices who it is intended should pass through the ITTU to employment outside. Stagnation is a danger leading to permanent employment in the ITTU and the blocking of opportunity for others. Consideration of these factors revealed a need for a training officer.

The training officer reports to the manager and is responsible to him for the operation of the technical training programmes of the engineering workshops of the ITTU. He has the status of a senior/principal/chief technician in the universities of Ghana.

The major concerns of the training officer are:

- ITTU Technical Apprentice Training Programme (TATP)
- Visiting Apprentice Training Programme (VATP)

The TATP is a five-year apprenticeship which in technical content is similar to most other technical apprenticeships provided by industrial companies but is unusual in that it aims at eventual self-employment. The training officer is responsible for the recruitment of candidates and the convening of an apprentice selection committee of the following composition:

Manager	Chairman
Training officer	Member
Workshop supervisor	Member
Section leader/s	Member/s

The training officer is responsible in consultation with the workshop supervisor for scheduling the training programmes and the rotation of apprentices between the sections of the ITTU. He provides theoretical and practical instruction and invites the participation of section leaders, consultants and clients with special skills. He carries out annual assessment tests of all apprentices and advises the manager on the promotion of successful apprentices and the termination of unsuccessful ones. During the final two years of the apprenticeship he arranges instruction in basic bookkeeping, accounting and business management to equip the apprentices for eventual self-employment.

The VATP is a programme of tailor-made, on-the-job training courses for individual clients of the ITTU who seek to acquire a new skill. The training officer consults with the client, the manager and the workshop supervisor to arrange a schedule that meets with the needs of the client and of the ITTU. The visiting apprentice usually attends the ITTU for less than two years and often on a part-time basis. The training officer ensures that the necessary theoretical and practical instruction is given and that the apprentice acquires the skill that he needs.

The success of the training officer's role depends essentially upon his relationship with the workshop supervisor. There must be flexibility on both sides to ensure that their objectives are achieved in harmony. Good training leads to high work output in quantity and quality. In the short term the needs of training may delay production but in the longer term production benefits. Both the training officer and the workshop supervisor must be prepared to balance the long-term and short-term needs to ensure the success of both the production programme and the training programme.

Training is a central concern of the ITTU. Although responding to the needs of an informal setting, an entirely informal approach to training will not suffice in achieving the objectives of technology transfer. The experience gained at an ITTU is in a way a transmutation from an informal to a formal mode of thought and activity. Some formal structure must be imparted to the learning process and, above all, progress must be monitored to ensure forward movement. This whole area is the concern and the responsibility of the training officer. Transferring technology to its clients is vital to the success of the ITTU.

Section leaders

Section leaders in the production workshops of the ITTU must adopt much the same approach, though on a smaller scale, as the workshop supervisor. One of the section leaders will also serve as the deputy or assistant workshop supervisor and in the usual course of events, a new workshop supervisor will in time arise from the ranks of the senior technicians in charge of a section. It is thus important that section leaders see their duties in the broader context of technology transfer and not in the narrow view of achieving only a high output of products. They must cooperate fully in the training programme and assist the training officer in providing a worthwhile on-the-job experience for apprentices assigned to their section.

The section leader of an engineering workshop in an ITTU reports to the workshop supervisor and through him to the manager, while the section leader of a non-engineering workshop in an ITTU reports to the rural and womens' industries extension officer and through him to the manager.

The section leader has the status of a technician/senior technician in the universities of Ghana, and is responsible for the work programme of his section. On the instructions of the workshop supervisor he undertakes the jobs assigned to his section ensuring that the work is carried out to specification and by the agreed completion date. He is also responsible for maintaining discipline in his section and for ensuring that all work is done by correct and safe procedures. He ensures that tools and safety equipment are correctly used and maintained and reports losses and breakages to the workshop supervisor.

Section leaders should be conscious of the fact that they, more than any other officers, actually determine the quality of the technology demonstrated, taught and transferred by the ITTU. Their role is a central one and essential to the success of the endeavour. The role of a section leader at an ITTU should provide more of a challenge and greater interest than that of a technician in most other institutions.

Rural and womens' industries extension officer

The rural and womens' industries extension officer (RAWIEO) reports to the manager and is repsonsible for the non-engineering projects and training programmes. He has the status of a senior/principal/chief technician in the universities of Ghana.

The RAWIEO maintains contact with local development agencies, Ghanian and foreign, and rural communities with a view to identifying new projects where the ITTU can manufacture equipment to upgrade an existing rural industry or introduce a new industry. All projects undertaken by the RAWIEO involve the ITTU in the manufacture of tools or machinery for supply to rural and urban, mostly informal, artisans.

The main areas of the RAWIEO's work are:

- Agriculture
- Food processing
- Building construction
- Rural crafts (textiles, pottery etc)

Many projects may be undertaken in collaboration with development agencies which provide funding and administration. The role of the ITTU may be limited to the manufacture of equipment or may include the provision of instructors. In some cases, it may be necessary for the ITTU to establish a production and training unit to provide longer-term training for the artisans. Such units may be operated permanently by the ITTU or administered by the collaborating agency with varying degrees of assistance from the ITTU.

In some projects, the training required is limited to a few days or a few weeks. In such cases the RAWIEO arranges short courses or workshops and these are often held in villages in the rural areas where the artisans will practise their crafts. On these occasions the RAWIEO takes part in the instruction, assisted as necessary by supporting staff or consultants from local or foreign institutions.

In maintaining contact with local communities the RAWIEO gives talks and video film presentations to popularize existing projects of the ITTU. He encourages the formation of craft associations to promote the interests of the artisans and to assist interaction with the ITTU.

The RAWIEO supervises the activities of the non-engineering production and training units of the ITTU and in this respect corresponds to the workshop supervisor with his responsibility for the engineering sections. The number of non-engineering sections is, however, normally fewer than the number of engineering sections.

Accounting officer

The accounting officer (AO) reports to the manager and is responsible for all the financial record keeping of the ITTU. He has the status of a senior/principal/chief accounting officer in the universities of Ghana.

The AO maintains books of accounts of all income and expenditure of the ITTU. He presents accounts of income and expenditure monthly and annually and he assists with annual and ad-hoc auditing of the ITTU accounts. He determines the costing of the products of the ITTU based upon data of material and labour content provided by the workshop supervisor. He supervises the work of the storekeeper and ensures that all stock control records are properly kept.

The manager is provided with all the accounting records needed for management purposes and for reporting to the Regional Advisory Board by the AO, who advises the manager on the cash position of the ITTU and its individual projects' funds and on the obligations in relation to taxation and statutory deductions from salaries.

The AO assists the training officer in providing instruction in basic bookkeeping for the ITTU technical apprentices, visiting apprentices and clients needing to acquire this basic skill.

Stores officer

The stores officer reports to the AO and through him to the manager of the ITTU. He normally holds the status of a technician in the universities of Ghana. He will have been trained as a mechanical engineering technician and will be familiar with the names of the tools used in the ITTU.

The stores officer is responsible to the AO for maintaining proper stock control records of the tools store, raw materials store and finished products store of the ITTU.

Of the three stores under his control the most active during working hours is the tools store and it is here that the stores officer stations himself during most of his duty hours. He issues tools and measuring instruments on the authority of the workshop supervisor, the training officer or a section leader. All items are signed for by the collector and are returned when the work is completed or at the end of the working day if the work continues into a second day. All tools kept in the tools store are returned to the store overnight and issued again the next day if needed to continue the work in hand.

In the case of materials, lubricants and consumable tools the stores officer maintains minimum stock levels and advises the accounting officer and the workshop supervisor when stocks fall to these levels so that replacements can be ordered in good time and before stocks become actually exhausted. The stores officer also advises his superior officers immediately of any losses or breakages that he detects in the course of his duties.

The stores officer is expected to keep the stores under his care in a neat and orderly manner. He should insist that every item returned to the store is clean and ready for re-use before returning it to its allotted place on the shelf where it can be found quickly when next it is needed.

The ITTU is a small organization and the stores officer normally works unaided or with, at most, one assistant. The efficient execution of his duties, however, can add much to the effectiveness of the producton and training programmes of the ITTU.

Security officer

The security officer reports to the manager and is responsible for the security of all the property of the ITTU including workshops and offices and their contents, vehicles and materials parked or stored externally on the ITTU compound and all property located at residential premises owned or leased by the ITTU. The security officer is usually appointed at a status equivalent to technician or accounting officer in the universities of Ghana. He will often be a retired police officer of the rank of sergeant or above. An essential feature of the work is the maintenance of good relations with the local police force and this relationship is facilitated by past employment in the force. The security officer assists the manager in recruiting suitable personnel to support his function. During working hours it is necessary to have on duty a security guard who is literate and can check documents and record the movements of people, vehicles and goods onto and off the compound. At night, it is necessary to have on duty three, four or five watchmen depending on the size, layout and location of the compound. Night-watchmen may mostly be illiterates who specialize in this work but the senior man should be literate and capable of taking any appropriate action in an emergency.

Other subordinate staff

There is perhaps little need to elaborate on the duties of other subordinate staff of the ITTU since they barely differ from those required by other institutions. It has however been found advantageous for senior staff to inform their subordinates fully about the aims and objectives of the ITTU. This not only adds interest to employment at subordinate levels but inspires a sense of purpose and service to the community that increases the efficiency of all operations. All subordinate staff should be encouraged to feel part of a team that is pioneering a new approach to the problem of grassroots industrial development. It is good for morale for all staff to feel that their work has some special significance and the resulting spirit pervading the ITTU will do much to convey its message and achieve its purpose.

Client selection and development

The client

Just as for a trading enterprise the interests of the customer must be paramount, for the ITTU the client is the most important person in the world. The ITTU exists to serve its clients and is designed primarily to cater for their needs.

The clients of the ITTU are the informal industrialists who gather together in informal industrial areas such as Suame Magazine in Kumasi. These men are popularly called 'fitters' in Ghana. They are artisans and craftsmen who practise many skills such as blacksmithing, welding, panel beating, carpentry etc. Some have undergone formal apprenticeships but many have been trained by masters who themselves operated in the informal sector. They are self-employed or have ambitions to be self-employed. Those already in business on their own account may work out of a concrete workshop, a wooden shack, or lean-to, or an inadequately stocked toolbox. Some possess more tools than others but all lack essential items of tooling and manufacturing equipment. If they employ others in their work it is usually in the capacity of unpaid apprentices. In general, they produce only to order, stock no materials or finished products and keep no records of their business.

The potential clientele of the ITTU may number several tens of thousands of artisans, though not all will approach the ITTU for help. Although the impact of the ITTU may reach all in time, this will result through copying and the spread of skills through the informal apprenticeship system. The ITTU is able to deal directly with only a small fraction of the total population of an informal industrial area such as Suame Magazine in Kumasi. It cannot even meet the needs of all who call upon it. Thus it is essential that some process of client selection is applied to ensure that help is given to the most able and the most entrepreneurial of the callers. However, before the selection process can be discussed it is necessary to look at the needs of the clients.

Needs of clients

The needs of informal industrialists can be divided into those needs perceived by the clients themselves, and needs which are more readily apparent to the adviser.

The perceived needs of clients are typified by the demands of the Suame Mechanical Association when the Suame ITTU opened in 1981. These were as follows:

1 An improved electricity supply to the workshops
2 Machine tools and equipment
3 Training for apprentices in skills which cannot be taught in existing informal workshops

The supply of these perceived needs is readily taken up by the artisans. There is no problem in a person accepting what he requested.

The apparent needs are clearly seen by the adviser but are not felt by the client. Some of these needs can be listed as follows:

1 Keeping of accounting and production records
2 Employing a systematic method of costing
3 Anticipating future demand by:
   (i) stocking raw materials
   (ii) producing goods for stock
4 Collaborating with other producers who have complementary manufacturing facilities so that all can produce a wider range of products in greater numbers and of better quality
5 Employing skilled labour as well as apprentices
6 Combining with others to bring pressure on government for the allocation of essential imported materials and tools

The client is often resistant to these apparent needs even when they are mentioned to him and so they cannot easily be supplied. They must usually be supplied indirectly or even covertly in the process of supplying the perceived needs. Thus the educative role of the ITTU is interwoven with the apparently simple process of giving the client what he asks for. However, meeting the client's perceived needs may exceed the resources of the ITTU, and so it is essential to employ a selection process to determine who will benefit most from the application of scarce resources.

Process of selection

When a client calls at the ITTU or is met at his own workshop he should be given the opportunity to describe his whole operation and to state his needs and difficulties as he sees them. Even at a first meeting it may be possible to supply some information or give some advice which will help the client. Remember that in Ghana all information has a value and it is not usual to give information or advice freely. Thus the free supply will encourage the client and let him feel that he has already benefitted from this contact with the ITTU.

There are other simple services which the ITTU can supply at no cost to itself. These include the sale of products made at the ITTU or the supply of engineering services, such as tool sharpening, for which an appropriate charge is made.

From the client's description of his needs it is usually apparent to the adviser that there are one or two things that the client could supply for himself. The appropriate action should be suggested to him. If subsequently he is found to have followed the advice it is a good indication that the client is serious in his enquiry and may have entrepreneurial qualities.

Everyone will stand in line for a hand-out. The ITTU must be presented as a means of helping those who make an effort to help themselves to overcome their own problems. Thus new clients must never be offered major assistance like the supply of a machine tool. Clients must prove their worth before receiving help of this kind. The process of selection to this level can extend typically to two years or more.

When a client is seen to be taking advice and comes back to the ITTU for further assistance he may be helped in more tangible ways. According to his need he may benefit from one or more of the following services:

- Training at the ITTU for himself or his employees
- Hiring of time on ITTU machines
- Sale of essential small tools such as lathe tool bits, taps, dies or drill bits
- Seconding of a technician to provide training or to commission a new production method in the client's workshop
- Assisting with the installation, commissioning or repair of a machine acquired by the client from his own resources.

When a client is found to have applied himself seriously to the adoption of a new product or process at his own workshop increasingly valuable services can be provided by the ITTU in promoting technology transfer. The client then becomes intimately associated with the ITTU and its group of leading clients who together form a production entity with closely integrated activities. This select group of clients may benefit from several of the following services:

- Subcontracting of orders placed on the ITTU. This may take the form of the supply of complete products or part manufacture of products to be completed at the ITTU (or by another client)
- Sharing of imported raw materials supplied by government import licence granted to the ITTU or through foreign aid projects
- Sale of new and used machine tools supplied through foreign aid projects
- Letting of workshop accommodation at the ITTU on a multi-year (typically three years) lease

One aim of the ITTU is to encourage a high degree of co-operation between clients who can benefit greatly from the services of others. The ITTU can often overcome natural distrust and prejudice by playing the role of honest broker until the barriers are overcome. The ITTU then becomes the centre of a matrix of manufacturing enterprises whose activities are so integrated that the impact of the whole is much greater than the sum of the parts. New clients come up and take their place in this matrix. However, they achieve this status only by completing an obstacle course, the object of which is to assess their capabilities and their capacity to learn. Throughout this process, the client's apparent needs are also supplied. As he becomes more intimately associated with the operation of the ITTU his methods become more formal in nature. He finds it necessary to keep records, to adopt the costing procedures employed at the ITTU, to stock or share in the stocking of raw materials, to produce finished goods for stock and generally to begin to think and act as a manufacturer in the modern sense of the word.

The informal artisan in his aboriginal setting tends to think like an informal sector trader. Contact with the ITTU can effect an evolution into a more rational, formal and productive thought process which is better adapted to the needs of industrialization. It is an evolution during which the individual grows in his sense of discipline and responsibility. It is necessarily a slow process and one which it is best not to hurry. ITTU staff need to cultivate patience as well as imagination to guide their clients along this path.

Regional advisory/management boards

It is the long term aim of the GRATIS Project to establish the ITTU as an autonomous institution within its region. To this end it seeks from the outset to work with a recognized group of people in the region and to consult the group at every stage of implementation and operation of the ITTU.

Several years before an ITTU is established it has become the practice for the TCC or GRATIS to identify what at this stage is called an interest group. The essential qualification for membership of an interest group is an interest in and understanding of the ITTU but care is taken to draw members from the following sectors of the community:

- Informal sector artisans
- Regional administration
- Local technical institutions
- Local development agencies

The interest group is asked to help find a suitable building to accommodate the ITTU and to assist in supporting its lease or purchase. It is also asked to help find residential accommodation for the principal officers of the ITTU. Another important function of the interest group is to increase local awareness of the benefits to be derived from the services of the ITTU, especially by the small-scale and informal industrialists of the region, who are encouraged to form an ITTU Clients Association and to elect members to represent them on the Regional Advisory Board of the ITTU.

Soon after the commissioning of a new ITTU, a regional advisory board is formed. This usually includes some members of the interest group but is more broadly based to ensure a wider representation of local interests. The regional advisory board normally contains about 12 members composed as follows:

1 Regional economic planning officer
2 Regional industrial promotion officer
3 Principal of polytechnic or technical institute

4 Representative of the Association of Ghana Industries
5 Representative of Ghana National Association of Garages
6 Representatives of National Council on Women and Development
7 and 8 Two representative of ITTU Clients Association
9 Local traditional ruler (village chief)
10 Representative of the Department of Rural Housing and Cottage Industries
11 ITTU manager
12 Representative of GRATIS Project

There is a degree of flexibility concerning the number and affiliation of members of the regional advisory board but the pattern has become set to the extent that variations seldom exceed two or three deviations from the above list.

The regional advisory board is required to meet at least four times a year and a two-monthly schedule is usually adopted. Members receive sitting allowances and expenses to encourage high attendance. A quorum is set at a simple majority which for a 12-member board is seven. The board elects one member to serve as chairman.

Once the regional advisory board is inaugurated, the ITTU manager is expected to follow its advice on all matter relating to local custom, practice and tradition as well as local development priorities. Between meetings of the board, the manager is expected to consult the chairman if any special function is planned or if any emergency arises, but the manager remains responsible to the GRATIS Project director for the administration and discipline of the ITTU.

After a period expected to be not less than three years, the regional advisory board can apply to take over the ITTU and reconstitute itself as the regional management board. Of the various factors that must be considered in this process, the most important in practical terms is the financial viability of the ITTU. Before it can be considered to be autonomous, an ITTU must be generating sufficient income to sustain its operations. The three principal sources of income are:

- Income earned from the sale of products and services
- Grants from development agencies
- Government subvention

Of these, the first is generated almost totally from within the region, the second is partly derived from agencies operating in the region and the last originates from outside the region but could be channelled through the regional administration. It should be the aim of the ITTU management to generate sufficient income substantially to fund recurrent expenditure and to establish a close collaboration with development agencies operating within the region to gain grants for capital developments and the extension of activities. Experience has shown that most development agencies are eager to work with the ITTU in the extension of their projects as it is often the only local source of practical help with manufacturing, installation and repair and with the provision of related training programmes. In general, it can be said that development at the local level requires three vital ingredients: funding, viable ideas and practical means of implementation. Often, it is found that development agencies have funds but are short of viable ideas and bereft of technical support. In supplying these deficiencies the ITTU gains additional income to support and extend its activities.

The last step in securing local financial autonomy is for the ITTU to arrange to receive any government subvention that it needs through the regional administration as part of the administrative costs of the region. This may come to the ITTU direct from the regional administration or through the good offices of the regional development programme.

A second important requirement in relation to gaining regional autonomy is local self-sufficiency in respect of technical information. To this end, the ITTU seeks to establish close links with one or more local technical institutions such as a university, polytechnic or technical institute from the outset. The first ITTU came into existence as an extension arm of the University of Science and Technology, Kumasi, Ghana's leading technological institution; establishing similarly close links with other technical institutions at a regional level is a vital element in maintaining the tradition of basing help for the informal sector upon the formal structure of technical education in the country. The informal sector may appear to make do with much that is second or third rate but the ITTU seeks to supply it with technical information and advice that is of the highest quality locally available. It does this to a large extent by drawing upon the resources of a local technical institution.

With the establishment of a regional management board and the transfer of control of the ITTU, the manager reports to the chairman of the board and is responsible to him for the administration and discipline of the ITTU.

The future

When the first edition of this handbook was written in 1986, the Suame ITTU remained the only one in full operation. The proposal for the GRATIS Project had been submitted to government but awaited approval. The proposal contained plans to establish an ITTU in the principal urban centres of all ten regions of Ghana but it was felt that the precise nature of the new ITTUs should be allowed to evolve.

Suame Magazine was, and is, by far the largest informal industrial area in Ghana. By the 1980s the technologies employed there included the extensive use of machine tools (at least 130 by 1984) including specialized types for engine overhauling such as crankshaft grinders and cylinder reborers. Both electric-arc and gas welding were well established in many hands and between them steel fabricators and carpenters had demonstrated their ability to construct almost any type of vehicle body or trailer including the largest types of articulated road transport vehicles.

The expertize of the Magazine attracted to it all those vehicles which could not be repaired elsewhere. So when a vehicle was declared to be terminally sick, it died at Suame and thereby contributed its corpse to the nation's largest engineering raw material stockpile. If half of the artisans strove to extend the lives of the living, the other half survived by recycling the carcases of the dead. The vast accumulation of scrap at Suame Magazine formed the basis on which a wide range of manufacturing industries built their activities. By the early 1980s Suame Magazine had become a vast hive of activity and a great economic force that attracted most of the major banks to build splendid new branches along the Suame Takwa road to serve the needs of an increasingly prosperous community.

The critics of the ITTU concept pointed out that it was difficult to conceive of any engineering project failing at Suame. The ITTU may have introduced semi-automatic lathes, gear hobbing and iron foundries but the Magazine was ready for these innovations and would probably have acquired them in a few years anyway. The success of the Suame ITTU, it was said, was due more to the natural dynamism of the Magazine than to any special merit of the ITTU.

These were difficult arguments to counter. The TCC technical officers who pioneered the Suame ITTU were the first to praise the talents and entrepreneurial zeal of the Suame artisans and even they at times doubted if they would find an equal degree of receptiveness elsewhere. For this reason, Tamale was chosen as the site of the second ITTU. Situated 400km from Kumasi in the less developed Sahelian north of Ghana, Tamale presented a vastly different situation. It confronted the ITTU with a smaller and far less industrialized urban setting which had much stronger links with the surrounding rural areas.

The few informal sector engineering artisans who settled in the light industrial area around the ITTU would scarcely have been noticed if they had joined one of the 40 wards of Suame Magazine. While these people would sit at the storm-centre of the ITTU's impact, it was clear that activities at Tamale would interact with a much wider variety of clients than at Suame. This had already been demonstrated by Frank Robertson who, from 1982, operated an Appropriate Technology Centre for the TCC in Tamale and who is remembered as the man who introduced the local manufacture of cotton spinning wheels. So the ITTU at Tamale built a work programme quite different from the one at Suame. It found an immediate need to supply the needs of farmers, food processors and rural craftsmen and seemed to be destined to maintain these manufacturing operations for some years before the local light engineering industry grew up to the point where it could take over.

Another major difference between Tamale and Suame was the emphasis on repair work. At Suame Magazine the artisans lived primarily by offering repair services and so the ITTU sought to promote manufacturing and was seldom called on to undertake repairs. At Tamale, however, it was found that most major vehicle and machinery repairs were sent to Kumasi and the arrival of the ITTU meant that this 400km trek was no longer necessary. Most of the facilities of the ITTU were unique in the Northern Region and access to these facilities for repair work could not be denied to the community. Once again, it was realized that these activities would inevitably continue for several years before they could be transferred to client workshops. It was even necessary for the Tamale ITTU to add crankshaft grinding and cylinder reboring facilities since these facilities did not exist north of Kumasi even though they were so well established in private workshops at Suame Magazine that they had never been a concern of the Suame ITTU.

From the experience of the Tamale ITTU it was clear that the technologies employed in the ITTU had a useful role to play even when they were significantly more advanced than those of its clients. Basic engineering manufacturing technologies are needed everywhere and the ITTU could provide an invaluable service to any community lacking them. It might be many years before even the initial range of technologies are transferred to the private sector but from the beginning this was facilitated by the demonstration effect of the ITTU.

It was the experience of operating the Tamale ITTU that finally persuaded GRATIS to abandon the concept of the secondary ITTU. The secondary ITTU had been conceived as an ITTU for less developed regions to address the needs of secondary, i.e. non-engineering, industries by introducing new tools and machines for agriculture, food processing and craft industries. It was conceived that a secondary ITTU would need only a few basic engineering facilities for the repair of items supplied to secondary industry clients. However, the heavy demand made upon all the engineering facilities of the Tamale ITTU led to the view that an ITTU on the Suame pattern with a full complement of machine tools and manufacturing facilities should be established in the capital town of every region. This view has been further reinforced by experience at Tema, Ho and Sunyani and at Cape Coast where the ITTU set out with less than the full complement of machine tools but experienced a constant demand to make up the deficiency.

Thus for application throughout Ghana, the equipment and range of technologies demonstrated by the ITTU has been standardized on the Suame pattern with a few additions from experience elsewhere, principally Tamale and Tema. The initial range of 20 basic engineering technologies may be listed as follows:

Metal Machining	1	Centre lathe turning
	2	Capstan (turret) lathe operations
	3	Milling machine/gear cutting
	4	Shaping machine operations
	5	Drilling machine operations
	6	Tool and cutter grinding
	7	Crankshaft grinding
	8	Cylinder reboring/honing
	9	Metal spraying
Fabrication	10	Electric-arc welding
	11	Gas welding
	12	Sheet metal working
	13	Metal spinning

48

Forging	14	Blacksmithing
	15	Power forging
Foundry	16	Ferrous casting
	17	Non-ferrous casting
	18	Heat treatment/case hardening
Woodworking	19	Pattern making
	20	Carpentry

The range is quite extensive and necessitates equipping the ITTU work-shops with a total of about 30 major items of equipment in the form of machine tools, welding sets, furnaces, etc. This represents an initial capital investment at 1992 prices of about $200 000 when some machines such as milling machines and crankshaft grinders are purchased as good, used machines to limit the cost.

At the point where six of the ten regions of Ghana are served by an ITTU and plans are well in hand to supply the other four, speculation has begun about the possibilities for further extension of the concept. This speculation has originated from within Ghana and from outside.

Dr Francis Acquah, the promoter of the GRATIS Project, often ex-pressed the desire to see ITTUs established at the district level when all ten regional centres had been supplied. Ghana's ten regions are subdivided into some 110 districts with each region containing, typically, between 10 and 13. In view of the numbers involved it is likely that Dr Acquah had in mind small workshops or Appropriate Technology centres such as pre-ceded the appearance of the ITTUs in Tamale and Bolgatanga. This being the case they would draw upon the resources of the regional ITTUs for most manufacturing and serious repair efforts. More recent thinking has suggested that the need could be largely served by well-equipped client workshops and to this end blacksmith training programmes were initiated in 1991 at the Tamale and Cape Coast ITTUs. The aim was to identify and train blacksmiths from all districts in the region with a view to assisting them in acquiring additional facilities, in order to establish at least one good repair workshop in each district. The more expensive items of equip-ment were sold on credit terms or hired to the artisans until they could accumulate the resources to purchase them.

The plan to upgrade blacksmiths' workshops should meet the immediate need in most districts. There are however a few districts that may qualify for a full ITTU or a somewhat scaled-down version, called a Rural Tech-nology Transfer Centre (RTTC) and resembling the old but unused con-cept of the secondary ITTU, for one or more of the following reasons:

- They are located far from the regional centre, such as the extreme north and the extreme south of the long narrow Volta Region;
- They are important centres for industrial development such as Tarkwa in the Western Region or Techiman in Brong-Ahafo Region, or
- They are too large to be fully served by the existing regional ITTU, such as the national capital Accra.

In addition to the planned ten regional ITTUs there could be a need for perhaps two or three additional full ITTUs at locations with urban and engineering industrial potential and 10 or 12 secondary ITTUs or RTTCs at more remote rural locations with predominantly non-engineering industrial potential.

An interest in establishing ITTUs outside Ghana has been expressed by international development agencies such as UNECA, UNIDO, ODA and CIDA, by African governments such as Kenya and Zambia and by organizations representing artisans and manufacturers such as the Jua Kali Association of Kenya and the Ugandan Manufacturers Association. In West Africa interest has been shown by CIDA in extending the ITTU to Benin and by USAID in extending the ITTU to Mali. If and when ITTUs are established in other African countries this handbook will serve their needs as it will serve the needs of the new ITTUs yet to be established in Ghana.

Appendix

Technologies and products introduced by the TCC/ITTU 1972–93
The lists given below provide examples of the sort of innovation that have been and can be introduced by an ITTU.

Technologies introduced and transferred by TCC/Suame ITTU 1972–86

1 Capstan (turret) lathe operation (1973). The semi-automatic production of steel bolts, nuts, studs, bushings etc, in batches of hundreds and a few thousands

2 Hot forgeing using dies (1973) for the production of steel bolt heads, notably coach or carriage bolts

3 Horizontal milling of hexagonal steel bars (1973) for bolt and nut production. This technology was needed because only round bars were available as raw material

4 Semi-automatic thread tapping (1974) using tapping attachment on drilling machine. This technique reduced tap breakages compared to nut threading on the capstan lathe

5 Tangential die head for bolt threading (1979) on a capstan lathe. This technique gave longer die life and greater flexibility of product range

6 Fly press stamping (1981) of steel washers from steel sheet cut-offs and broaching (1982) of splines in chain sprocket wheels

7 Tube bending (1981) usually locally-made bender based on vehicle wheel hub

8 Horizontal milling of gear forms (1982) for production of gear wheels and chain sprocket wheels

9 Gear hobbing (1983) for the production of gear wheels

10 Shaping machine operations (1983) for production of hammer heads and other sheet metal working tools

11 Tool and cutter grinding machine operations (1975) for the regrinding of milling cutters and threading dies

51

12 Electric furnace case hardening (1982) of chain sprocket wheels

13 Foundry operations for the production of castings:
   (a) Brass and bronze (1975)
   (b) Aluminium (1983)
   (c) Iron (1983)
   (d) Various alloys (1985) including hard cast iron and aluminium bronze

14 Sheet metal working techniques (1974) for rolling, folding, shearing, flame cutting and gas and electric welding

15 Refractory cement for furnace lining (1985) produced from local raw materials

Products manufactured by ITTUs and clients 1980–93
The products listed below were either newly introduced, greatly improved in quality or produced by a new method.

A   *Produced by metal machining*
 1 Steel bolts and nuts including:
   (a) Hexagon headed bolts (UN, BS and metric)
   (b) Coach or carriage bolts
   (c) Wheel bolts and nuts for vehicles
   (d) Spring centre bolts for vehicles
   (e) Studs
   (f) Tube inserts for tubular steel furniture
   (g) Sparking-plug extenders

 2 Steel washers

 3 Gear wheels

 4 Chain sprocket wheels

 5 Engine valve inserts for cylinder head reconditioning

 6 Engine cylinder liners

 7 Piston rings for vehicle engines and gas compressors

 8 Bearing bushings, sleeves and shells

B   *Produced by hand forgeing*
 9 Hoes

10 Cutlasses

11 Hammer heads

12 Panel-beating tools

13 Rubber tapping knives

14 Gate hinges

15 Coach bolt heads

C *Produced by casting*

16 Bar stock of brass, bronze, light alloy and iron produced from scrap

17 Corn mill and pepper mill grinding plates

18 Diesel engine cylinder liners

19 Bullock plough blades

20 Corn mill bearing shells

21 Light alloy pistons for vehicle engines

22 Charcoal-burning cooking stoves (coal pots)

23 Bronze water pump impellers

D *Produced by welding and steel fabrication*

24 Sheet steel tanks for:
   (a) Soapmaking
   (b) Palm oil production
   (c) Storage

25 Presses for:
   (a) Palm oil production
   (b) Gari (from cassava) production

26 Carpenter's saw bench (table saw)

27 Wood turning lathe

28 Block making machines for
   (a) Concrete blocks
   (b) Soil-cement blocks

29 Corn milling machines

30 Animal feed milling machine

31 Cassava grating machines

32 Palm fruit pounding machines

33 Palm kernel cracking machines

34 Seed planters for maize and cowpeas

35 Rice threshers with pedal drive

36 Hulling machines for maize and rice

37 Sub-assemblies for bullock ploughs and bullock carts

38 Sawdust-burning cooking stoves

39 Smokers for beekeepers

40 Wood-planing machine

41 Bicycle trailer

42 Small farm vehicles (wheelbarrows)

43 Power hammers

44 Winches for well sinking

45 Bench-mounted shears

46 Sheet metal-folding machine

47 Lift-out crucible furnaces for iron casting

48 Tilting furnaces for iron casting

49 Metal spinning lathe

E   Produced by carpentry

50 Soapmaking equipment such as:
    (a) Cutting tables
    (b) Moulding boxes
    (c) Cartons

51 Weaving looms for broadcloth

52 Beehives

53 Bottled drink crates

54 Turned wooden products such as:
    (a) Furniture legs and bars
    (b) Bowls and lids
    (c) Pestles (Ata)
    (d) Candleholders

55 Cotton-spinning wheels

56 Patterns for foundry products

57 Solar beeswax extractors

www.ingramcontent.com/pod-product-compliance
Ingram Content Group UK Ltd.
Pitfield, Milton Keynes, MK11 3LW, UK
UKHW021815300426

5500IPUK00004B/31